Notice to Students

There are 318 questions, which are quite extensive for each topic. Every other question is explained to make each topic of anatomy and physiology easy to comprehend. Understanding these questions can play a helpful role in getting the grade you strive for. Make sure to also purchase the following at SEPBOOKS.COM

ANATOMY AND PHYSIOLOGY STUDY GUIDE: *KEY REVIEW QUESTIONS AND ANSWERS WITH EXPLANATIONS*
(Topics: Orientation of the Body, Structure and Function of Cells, Structure and Function of Tissues, Integumentary System) Volume 1

ANATOMY AND PHYSIOLOGY STUDY GUIDE: *KEY REVIEW QUESTIONS AND ANSWERS WITH EXPLANATIONS*
(Topics: Bone Tissue, Skeletal System, Muscle Tissue, Muscular System) Volume 2

ANATOMY AND PHYSIOLOGY STUDY GUIDE: *KEY REVIEW QUESTIONS AND ANSWERS*
*(Topics: Cardiovascular System, Endocrine System, Lymphatic System) Volume 4 *This book will be available in Fall of 2010**

Contents

Anatomy and Physiology
Nerve Tissue

1. The central nervous system is composed of which of the following?
    a. spinal cord and muscles
    b. muscles and brain
    c. bones and muscles
* d. spinal cord and brain

The nervous system is made up of the peripheral and central nervous systems. The peripheral nervous system (PNS) houses the nervous tissue that is outside the central nervous system (CNS). The PNS is divided into the automatic nervous system and the somatic nervous system.

2. Which of the following is not a component of the peripheral nervous system?
    a. peripheral nerves
    b. spinal nerves
* c. spinal cord
    d. cranial nerves

3. Which of the following is a star-shaped cell that is important for the production of nerve impulses?
    a. microglia
* b. astrocytes
    c. ependymal cells
    d. oligodendrocytes

Microglia are key cells that destroy bacteria in the body. They have a key function in immunity. Ependymal cells are made of columnar and squamous epithelium; they function in lining the ventricles of the brain. Cerebral spinal fluid is created in the ventricles of the brain. Oligdendrocytes form the myelin sheath, which is found around the axons of neurons. Astrocytes also connect blood vessels and neurons. Microglia, ependymal cells, oligodendrocytes and astrocytes are all found in the CNS.

4. Which are the most common types of neuroglia of the CNS that have fewer processes than an astrocyte?
* a. oligodendrocytes
    b. microglia

c. satellite cells
d. ependymal cells

5. Which of the following create myelin sheaths around axons in the peripheral nervous system (PNS)?
a. microglia
* b. Schwann cells
c. satellite cells
d. oligodendrocytes

Schwann cells (neurolemmocytes) and satellite cells are both found in the PNS. Satellite cells have a supportive function around the perimeter of cell bodies of neurons. Microglia in the CNS destroy bacteria and invading microbes by the process of phagocytosis.

6. Which is the most common type of neuron in the body?
a. efferent
b. afferent
* c. association
d. axolemma

7. A space on an axon where there is an absence of myelin is referred to as:
* a. node of Ravier
b. dendrite
c. cell body
d. Schwann cell

Neurons conduct nerve impulses in the body. The neuron contains dendrites, a cell body, axons, nodes of Ravier, and synaptic end-bulbs. Dendrites are branches of a neuron that attach to the cell body. Dendrites conduct nerve impulses toward the cell body. The cell body contains organelles like the following: mitochondria, nucleus, lysosomes, mitochondria and a cytoplasm. The axon is connected to the cell body; it has the appearance of a long tube. Most of the axon is covered by a fat covering called a myelin sheath. The end of the axon has structures that are called synaptic end-bulbs. Synaptic end-bulbs are utilized as a storage space for neurotransmitters.

8. Which is the region where the cell body connects to the axon?
a. synaptic end-bulb

b. dendrite
c. nucleus
* d. axon hillock

9. All of the following statements regarding myelin are true except:
a. Myelin increases the speed of a nerve impulse.
b. Increased speed of muscle contraction in an adult is due to having more myelin than a child.
* c. If an axon did not have a myelin sheath, a nerve impulse would be impossible.
d. Oligodendrocytes create myelin in the CNS.

A nerve impulse is possible if a myelin sheath was not present around an axon. The nerve impulse would just be much less in speed compared to a myelinated axon. Axons without a myelin sheath are considered unmyelinated.

10. Which of the following is not found in the cell body of a neuron?
a. lysosome
b. mitochondria
* c. axon
d. golgi complex

11. Which of the following are sensory in nature, conduct nerve impulses from the muscles and then to the brain?
* a. afferent neurons
b. efferent neurons
c. axon hillock
d. ependymal cells

Efferent neurons conduct motor nerve impulses from the spinal cord and brain to the muscles. There is another functional type of neuron called an association neuron. This type helps in bringing one nerve impulse from one neuron to another neuron. In this way, they aid in connecting the impulse message from neuron to neuron.

12. Which physical type of neuron has one cell body, an axon and many dendrites?
a. unipolar neuron
b. bipolar neuron

* c. multipolar neuron
d. synaptic end-bulb

13. Which type of neuron has only one process that is connected to the cell body?
a. bipolar
* b. unipolar
c. multipolar
d. synaptic end-bulb

Both multipolar and bipolar neurons have a cell body that physically separates the axon from the dendrites. Multipolar neurons are most abundant in the central nervous system (CNS). The synaptic end-bulbs store neurotransmitters, which are freed at the synapse.

14. The perikaryon is identified as which of the following in a neuron?
a. dendrite
b. axon
c. node of Ranvier
* d. cell body

15. White matter is _____________ and gray matter is ____________. The white matter surrounds the gray matter in the spinal cord.
* a. myelinated, unmyelinated
b. unmyelinated, myelinated
c. the cell body, the dendrite
d. the dendrite, the cell body

White matter contains myelin, which is made of a fat lining that conducts nerve impulses at a faster rate. White matter contains axons with myelin coating that is found in the CNS. Gray matter doesn't have myelin. Gray matter contains dendrites, neuroglia, cell bodies and unmyelinated axons.

16. Which of the following types of neuroglia serve as a macrophage?
a. satellite cell
* b. microgila
c. Schwann cell
d. astrocyte

17. A multipolar neuron has more than one of which of the following

structures?

a. cell body
b. axon
* c. dendrite
d. soma

Multipolar neurons have one axon and many dendrites. Soma is another word for cell body.

18. Which type of nerve fiber is myelinated, has the largest diameter and conducts nerve impulses at the greatest speed?

* a. A fiber
b. B fiber
c. C fiber
d. axolemma

19. Which type of nerve fiber is not myelinated, has the smallest diameter and conducts impulses at the lowest speed?

a. A fiber
b. B fiber
* c. C fiber
d. synaptic end-bulb

Both A and B nerve fibers are myelinated. Type B nerve fiber conducts impulses usually at a rate that is slower than Type A nerve fiber but faster than type C nerve fiber.

20. Which of the following is a type of nerve cell that conducts impulses?

* a. neuron
b. myelin
c. organelle
d. action potential

21. An action potential can be most closely described as which of the following?

a. nerve cell
* b. nerve impulse
c. ion
d. white matter

The membrane around the neuron called the plasma membrane has little

holes for ions to enter. The positive ions that enter can be potassium, calcium or sodium. The negatively charged ions inside the plasma membrane of the neuron attract the positive ions. This action will soon lead to an action potential. Before the ions entered the plasma membrane, the voltage inside the neuron was at the resting membrane potential, which was -70 mV. However, when the positive ions entered, the voltage inside then went up to about +3 mV as an example. The voltage goes back again to -70 mV. The changing from -70 mV to +3 mV and then back again to -70 mV makes up when the action potential or nerve impulse occurs.

22. The plasma membrane of a neuron changes from -65 mV to +15 mV. This most closely describes which of the following?
* a. depolarization
  b. hyperpolarization
  c. all-or-nothing principle
  d. repolarization

23. Which of the following values would point to hyperpolarization of a plasma membrane of a neuron if the resting membrane potential was originally -68 mV?
  a. -15 mV
  b. +15 mV
  c. +85 mV
* d. -85 mV

The value +15 mV would point to a depolarization of the plasma membrane of a neuron. This means that the voltage goes from a negative value to a more positive value. For an action potential to occur the plasma membrane must reach a certain value, which is called the threshold potential.

24. ____________ usually begins right after the voltage of the plasma membrane of a neuron reaches a positive peak. This will eventually lead to the voltage to be brought back to the original resting membrane potential.
  a. Depolarization
* b. Repolarization
  c. Threshold potential
  d. All-or-none principle

25. If the original resting membrane potential was -60 mV and the action

potential peaked at +25 mV, which value would most closely represent the beginning of depolarization?

a. -60 mV

b. -55 mV

* c. +25 mV

d. 0 mV

Both depolarization stage and repolarization stage make up a total action potential. There is a certain time period that another action period cannot be produced after the first action potential was created. This is referred to as the refractory period.

26. An action potential either occurs fully or does not occur. This describes which of the following?

a. hyperpolarization

* b. all-or-none principle

c. threshold

d. depolarization

27. Continuous conduction usually occurs with ___________fibers of axons, and saltatory conduction usually occurs with _________ fibers of axons.

a. myelinated, myelinated

b. unmyelinated, unmyelinated

c. myelinated, unmyelinated

* d. unmyelinated, myelinated

Saltatory conduction makes the nerve impulse much faster than continuous conduction. In saltatory conduction the nerve impulse goes from one node of Ranvier to the next node of Ranvier. This is like skipping portions of the axon to get the action potential to occur faster. In continuous conduction the nerve impulse goes across the entire length of the unmyelinated axon. This is a much longer process.

28. Which type of nerve fiber has the longest absolute refractory period?

a. type A fiber

b. type B fiber

* c. type C fiber

d. nucleus

29. Which of the following is a junction or can be described as small little bridges between neurons?

a. dendrites
b. axons
* c. synapses
d. action potentials

There are chemical synapses which are the slower kind of synapses. This kind of synapse doesn't have small bridges or connections between neurons. Instead, it has a synaptic cleft, which is an opening between neurons with no bridge or junction. The synaptic cleft has chemicals called neurotransmitters that have to travel like a ship in water from one neuron to another. In this way, nerve conduction is possible. Electrical synapses permit faster communications between neurons. Thus, the nerve impulse will be faster. An electrical synapse has a junction or bridge that connects one neuron to the next allowing for faster communication.

30. Which of the following is not a structure found on an axon?

* a. dendrite
b. neurolemma
c. axolemma
d. node of Ranvier

31. Gray matter can be found in which of the following areas of the body?

a. legs and arms
* b. spinal cord and brain
c. brain and arms
d. spinal cord and legs

Gray matter doesn't contain myelinated tissue. It is found in the CNS which consists of the spinal cord and brain.

32. Which of the following utilizes neurotransmitters for nerve conduction?

a. electrical synapse
b. all-or-none principle
c. junction
* d. chemical synapse

33. Which of the following is a type of neurotransmitter that is discharged by neurons in the CNS and PNS?

* a. acetylcholine (ACh)
  b. axon
  c. glutamate
  d. soma

Acetylcholine can slow the speed of the heart beat. It is also an inhibitory and excitatory neurotransmitter. Neurotransmitters are expelled from the synaptic end-bulbs.

34. Efferent neurons conduct ____________ nerve impulses from the spinal cord and brain to the glands and ___________.

  a. sensory, muscles
* b. motor, muscles
  c. motor, spinal nerves
  d. sensory, spinal nerves

35. Afferent neurons are sensory and conduct nerve impulses from the CNS to the PNS.

  a. True
* b. False

Efferent neurons conduct nerve impulses from the CNS to the PNS. Afferent neurons conduct nerve impulse from the PNS to the CNS. This means that sensory nerve impulses are conducted from the muscles and skin to the spinal cord and brain.

36. Pertaining to a chemical synapse, which of the following divides the postsynaptic and presynaptic neurons?

  a. dendrite
  b. axon
  c. gap junction
* d. synaptic cleft

Spinal Nerves & Spinal Cord

1. Which of the following is one of the meninges that makes up the outer covering of the brain and spinal cord?

a. pia mater
b. arachnoid
c. conus medullaris
* d. dura mater

The pia mater, arachnoid and dura matter are the three types of meninges, which are found around the spinal cord and brain. The pia mater makes up the inner covering of the meninges. The arachnoid is the middle covering of the meninges. The conus medullaris is the portion at the end of the spinal cord where it becomes like a point of a cone.

2. Which one of the meninges is spider-like in appearance and is found between the inner and outer coverings of the brain and spinal cord?

a. pia mater
b. dura mater
* c. arachnoid
d. conus medullaris

3. The spinal cord starts superiorly at the ____________ and ends inferiorly at approximately ____________.

a. pons, T12
* b. medulla, L2
c. midbrain, L2
d. medulla, T12

The lowest portion of the brain is the medulla. The spinal cord varies in diameter throughout the spine. There is a lumbar enlargement of the spinal cord at T9 to about T12. There is also a cervical enlargement of the spinal cord at C4 to about T1.

4. Since there are seven cervical vertebrae in the normal adult human spine, there are ____________ cervical pairs of spinal nerves.

a. 5
b. 6
c. 7
* d. 8

5. Which of the following contains cerebral spinal fluid, which is essential for health of the nervous system?

a. pia mater
b. dura mater
* c. subarachnoid space
d. arachnoid

Cerebral spinal fluid or CSF protects the spinal cord since it acts as a shock absorber. CSF is secreted by the ventricles of the brain. The subarachnoid space contains spinal arteries and veins.

6. Which of the following separates the arachnoid and the dura mater?

a. pia mater
b. denticulate ligaments
c. subarachnoid space
* d. subdural space

7. The denticulate ligaments are structures composed of which section of the meninges?

* a. pia mater
b. arachnoid
c. dura mater
d. subarachnoid space

Denticulate ligaments help to support and protect the spinal cord. They are located between L1 nerve root and the foramen magnum.

8. The cervical enlargement is located in which area of the spinal cord?

a. C1-C2
b. C2-C5
* c. C4-T1
d. C5-T4

9. Which of the following is a false statement?

a. There are 5 pairs of sacral nerves.
b. There are 5 pairs of lumbar nerves.
* c. There are 31 spinal nerves in the spine.
d. There are 12 pairs of thoracic nerves.

There are 31 spinal nerves on each side of the spine. In other words, there are 31 pairs of spinal nerves. There are also 8 pairs of cervical nerves and

one pair of coccyx nerves. However, some people don't have this one pair of coccyx nerves. For anatomical purposes remember that there are 62 spinal nerves or 31 pairs of spinal nerves.

10. The two halves of the gray matter are connected by a structure called the:

a. anterior white commissure
* b. gray commissure
c. anterior median fissure
d. posterior root ganglion

11. The central canal is a structure found in which of the following locations?

a. anterior gray horn
b. anterior white commissure
* c. gray commissure
d. posterior median sulcus

Anterior gray horns, lateral gray horns, gray commissure and central canal are all structures of gray matter in the spinal cord. Posterior median sulcus and anterior white commissure are structures found in the white matter of the spinal cord.

12. Which type of tracts in the spinal cord carries nerve impulses to the brain?

a. cervical plexus
b. descending
c. motor
* d. sensory

13. Spinal nerve roots extended at the end of the spinal cord are referred to as which of the following?

a. filum terminale
* b. cauda equina
c. lumbar enlargement
d. conus medullaris

The order from superior to inferior of the above structures is the following: lumbar enlargement, conus medullaris, cauda equina and filum terminale.

14. The filum terminale is a type of tissue located at the coccyx and is composed of:
* a. pia mater
b. dura mater
c. arachnoid
d. ventricle

15. The connective tissue covering directly around a bundle of nerve fibers of a spinal nerve is most accurately termed:
a. endoneurium
b. epineurium
c. fascicle
* d. perineurium

A bundle of nerve fibers of a spinal nerve is called a fascicle. The whole nerve is surrounded by a connective tissue sheath referred to as the epineurium. Each nerve fiber is wrapped by a connective tissue shealth called the endoneurium.

16. The dorsal root of a spinal nerve carries _____________ fibers, which are ____________. Mixed nerves are considered spinal nerves.
* a. afferent, sensory
b. efferent, sensory
c. afferent, motor
d. efferent, motor

17. The entire spinal nerve is surrounded by which of the following connective tissue sheaths?
a. fascicle
* b. epineurium
c. perineurium
d. endoneurium

Spinal nerves are found in the PNS. The anterior root of a spinal nerve is motor and sends messages from the spinal cord to the muscles. The anterior root of a spinal nerve carries sensory fibers, which are afferent, to the dorsal horn of the gray matter of the spinal cord. For example, someone accidentally burns his finger on a hot pan. The sensory nerve impulse then travels from the finger and ultimately to the anterior root of the spinal nerve. From the anterior root this information (the pain) goes to the dorsal horn of

the gray matter of the spinal cord.

18. In which part of the spine does the ventral and dorsal nerve roots connect to form a spinal nerve?

a. intervertebral disc
b. central canal
* c. intervertebral foramen (IVF)
d. spinous process

19. All of the following are nerves that are branches of the cervical plexus except:

a. phrenic
b. supraclavicular
c. ansa cervicalis
* d. long thoracic

The long thoracic nerve is a branch found in the brachial plexus. The cervical plexus is made up of the merging ventral rami of cervical spinal nerves C1-C4. There are three different branches in the cervical plexus. There is a mixed branch, a muscular branch and a cutaneous branch. The phrenic nerve innervates the diaphragm; this nerve is extremely important for proper breathing. The phrenic nerve is a mixed branch of the cervical plexus. The ansa cervicalis is part of the muscular branch of the cervical plexus. It has an inferior and superior root. The supraclavicular nerve is one of the cutaneous branches of the cervical plexus.

20. Each nerve fiber of a specific spinal nerve is directly surrounded by which of the following?

* a. endoneurium
b. epineurium
c. perineurium
d. fascicle

21. Which of the following is a nerve located in the brachial plexus?

a. great auricular
* b. ulnar
c. supraclavicular
d. transverse cervical

The brachial plexus is made up of the merging of ventral rami of spinal

nerves C5-C8 and T1. The great auricular, transverse cervical and supraclavicular nerves are branches found in the cervical plexus.

22. All of the following are ventral rami origins of the phrenic nerve except:
* a. C2
b. C3
c. C4
d. C5

23. Which of the following is most posterior to the anterior median fissure on the spinal cord?
a. spinal nerve
b. gray commissure
* c. posterior median sulcus
d. lateral gray horn

Although the gray commissure is posterior to the anterior median fissure, the posterior median sulcus is most posterior. The gray commissure connects both halves of the gray matter. The central canal is located in the gray commissure. This canal is found throughout the spinal cord.

24. How many pairs of lumbar spinal nerves are there in the human spine?
a. 4
* b. 5
c. 6
d. 8

25. The dorsal scapular nerve arises from which spinal nerve root of the brachial plexus?
* a. C5
b. C6
c. C7
d. C8

The main branches of the brachial plexus are the axillary nerve, radial nerve, musculocutaneous nerve, median nerve and ulnar nerve.

26. The median nerve is a main branch of the brachial plexus; it arises from the ventral rami of spinal nerves _________.

a. C5
b. C5-C7
* c. C5-T1
d. C3-C5

27. The deltoid is innervated by which nerve arising from the brachial plexus?
a. median
b. radial
c. ulnar
* d. axillary

The axillary nerve also innervates the teres minor muscle. Remember that the median, radial, ulnar and axillary nerves not only innervate muscles but the skin around these muscles. The flexor carpi ulnaris is innervated by the ulnar nerve. The median nerve innervates flexor muscles of the forearm. The radial nerve innervates muscles that extend the wrist.

28. The biceps and brachialis muscles are innervated by which nerve arising from the brachial plexus?
a. median
b. ulnar
c. radial
* d. musculocutaneous

29. Which of the following is not a nerve arising from the brachial plexus?
a. thoracodorsal
* b. supraclavicular
c. musculocutaneous
d. lateral pectoral

The supraclavicular nerve is a cutaneous branch of the cervical plexus. It has an origin of C3-C4 ventral rami of cervical nerves. The thoracodorsal nerve has an origin of C6-C8 in the brachial plexus. The musculocutaneous nerve has an origin of C5-C7 in the brachial plexus. The lateral pectoral nerve has an origin of C5-C7 in the brachial plexus.

30. Which of the following nerves is not a main branch of the brachial plexus?
a. axillary

b. median
c. radial
* d. phrenic

31. Which of the following ventral rami of spinal nerves do not form into either a cervical, brachial, lumbar, or sacral plexus?
a. C1-C4
b. C5-T1
* c. T2-T12
d. L1-L4

The ventral rami of spinal nerves T2-T12 are identified as intercostal nerves. They exactly supply the muscles they innervate. The ventral rami of all other spinal nerves form plexuses. Ventral rami of spinal nerves L1-L4 make up the lumbar plexus.

32. Which of the following is not a nerve located in the lumbar plexus?
a. femoral nerve
b. obturator nerve
c. iliohypogastric nerve
* d. inferior gluteal nerve

33. The dorsal ramus of a spinal nerve would give nerve supply to which of the following locations?
a. skin and muscles of anterior chest
b. muscles of lower legs
* c. skin and muscles of posterior trunk
d. muscles of the feet

The ventral ramus of a spinal nerve gives nerve supply to the legs, arms and lateral and medial portions of the trunk. A spinal nerve splits into two different branches or rami. One is called the dorsal ramus and the other is called the ventral ramus.

34. Which is a nerve of the lumbar plexus that innervates muscles of the thigh that extend the leg?
a. iliohypogastric nerve
* b. femoral nerve
c. ilioinguinal nerve
d. ulnar nerve

35. Which nerve of the lumbar plexus is derived from ventral rami of spinal nerves L2, L3 and L4?

a. genitofemoral
b. iliohypogastric
c. ilioinguinal
* d. obturator

The genitofemoral nerve is derived from ventral rami of spinal nerves L1 and L2 of the lumbar plexus. The ilioinguinal and iliohypogastric nerves are derived from ventral ramus of spinal nerve L1 in the lumbar plexus.

36. A network of nerves is most closely referred to as a:

* a. plexus
b. dorsal ramus
c. ventral ramus
d. mixed nerve

37. Which nerve of the sacral plexus is derived from ventral rami of spinal nerves L4, L5, S1, S2 and S3?

a. nerve to obturator internis
b. inferior gluteal nerve
* c. sciatic nerve
d. superior gluteal nerve

The nerve to the obturator internis is derived from ventral rami of spinal nerves L5-S2. This nerve innervates obturator internis and superior gemellus muscles. The inferior gluteal nerve is derived from ventral rami of spinal nerves L5-S2. This nerve innervates the gluteus maximus muscle. The superior gluteal nerve is derived from ventral rami of spinal nerves L4-S1. This nerve innervates the gluteus minimus and medius muscles. The sciatic nerve forms into two branches, which are the tibial nerve and the common peroneal. The tibial nerve further divides into lateral and medial plantar nerves. The common peroneal nerve further divides into deep and superficial peroneal nerves.

38. The __________ nerve is derived from ventral rami of spinal nerves L5-S2 and is part of the ___________ plexus.

a. posterior femoral cutaneous, sacral
* b. inferior gluteal, sacral
c. obturator nerve, lumbar

d. femoral nerve, lumbar

39. Which of the following is not a nerve of the sacral plexus?

a. inferior gluteal
b. superior gluteal
* c. genitofemoral nerve
d. sciatic nerve

Remember that each plexus is found on both sides of the body. There are two cervical plexuses, two brachial plexuses, two lumbar plexuses and two sacral plexuses. Other nerves of the lumbar plexus include the iliohypogastric, ilioinguinal, genitofemoral, obturator, femoral, and lateral femoral cutaneous nerves. Nerves of the sacral plexus include the superior gluteal, inferior gluteal, perforating cutaneous, posterior femoral cutaneous, sciatic, tibial, common peroneal and pudendal nerves. Nerves to the piriformis, quadratus internus, superior gemellus and obturator internus are also nerves of the sacral plexus.

40. The largest nerve in the body is considered the __________ nerve. It is located in the __________ plexus.

* a. sciatic, sacral
b. sciatic, lumbar
c. femoral, sacral
d. femoral, lumbar

41. Which is a branch of the sciatic nerve that innervates the hamstring muscles?

a. femoral nerve
b. medial plantar nerve
c. superior gluteal
* d. tibial nerve

The femoral nerve can be eliminated as a possible choice because it is part of the lumbar plexus. The medial plantar nerve is a branch of the tibial nerve. However, the medial plantar nerve innervates muscles of the foot. The superior gluteal nerve is a nerve of the sacral plexus, and it is not a branch of the sciatic nerve.

42. The supraclavicular nerve is located in which plexus of the body?

a. brachial

*   b. cervical
         c. lumbar
         d. sacral

43. All of the following are related with automatic reflexes except:
*   a. skeletal muscle
         b. smooth muscle
         c. glands
         d. cardiac muscle

Somatic reflexes are related with skeletal muscle contraction. An example of a monosynaptic reflex is the stretch reflex. The stretch reflex consists of one synapse, one sensory neuron and one motor neuron.

44. The femoral and obturator nerves are found in which region of the body?
         a. cervical plexus
         b. brachial plexus
         c. sacral plexus
*   d. lumbar plexus

45. All of the following are true regarding polysynaptic reflex arcs except:
*   a. There is only one type of neuron involved.
         b. There are more than two types of neurons involved.
         c. There are two or more synapses involved.
         d. There can be three or more synapses involved.

Monosynaptic reflex arcs involve just one synapse. There are five parts of a reflex arc which includes the effector, motor neuron, sensory neuron, spinal integrating center and receptor. The integrating center has interneurons that link sensory neurons to motor neurons. The effector is considered the gland or muscle. It produces a response or reflex. The path of the reflex arc first begins with some kind of stimulus. For example, the stimulus will be the pain of touching a hot stove and is communicated to the receptor, which is part of the sensory neuron. This message is communicated to the sensory neuron axon terminals. Then this message goes to the spinal integrating center. This message next travels to the motor neuron. The message is then sent to the muscles of the fingers (the effector) to withdraw away from the hot stove.

46. All of the following are nerves derived from ventral rami of spinal nerves of the brachial plexus except:

a. radial
b. ulnar
* c. transverse cervical
d. median

47. Which of the following plexuses have a nerve supply to the muscles of the legs?

a. lumbar, cervical
b. cervical, brachial
c. sacral, brachial
* d. lumbar, sacral

The cervical plexus has a nerve supply to the diaphragm, muscles of the shoulders and neck. The brachial plexus has a nerve supply to the muscles of the shoulders, neck and arms.

48. Which of the following is a part of the reflex arc that is responsible for reacting to a motor nerve impulse?

* a. effector
b. receptor
c. integrating center
d. sensory neuron

49. Which bulging structure encloses the cell bodies of sensory neurons?

a. ventral root of a spinal nerve
b. cranial nerve
* c. dorsal root ganglion
d. posterior gray horn

There are 12 cranial nerves. Some functions of these nerves include the following: taste, facial expression, hearing, sight, equilibrium and smell.

50. Which tract or tracts of the spinal cord is for touch, deep pressure and distinguishing pain and temperature?

a. lateral corticospinal tracts
* b. spinothalamic tract
c. rubrospinal tract
d. tectospinal tract

Cranial Nerves & Brain

1. All of the following are part of the brainstem except:
    a. pons
    b. medulla oblongata
*   c. ventricles
    d. midbrain

The brain stem from inferior to superior is the following: medulla oblongata, pons and then the midbrain. There are four ventricles where cerebrospinal fluid can be found. The four ventricles are located in the brain. There are two lateral ventricles, a third ventricle and the fourth ventricle which is most inferior.

2. Which of the following is identified as the forebrain?
    a. rhombencephalon
    b. mesencephalon
*   c. prosencephalon
    d. diencephalon

3. Which region of the brain is directly above the brainstem and consists of the thalamus and hypothalamus?
*   a. diencephalon
    b. lateral ventricle
    c. fourth ventricle
    d. cranial meninges

The cranial meninges consist of the dura mater, archnoid and pia mater. Each lateral ventricle is found in each side of the brain. The majority of the brain is composed of the cerebrum, which has a left and right hemisphere.

4. Which of the following is not a location for the circulation of cerebrospinal fluid?
    a. lateral ventricles
    b. subarachnoid space
    c. around spinal cord
*   d. spinal nerve

5. Cerebrospinal fluid is formed in which of the following specific locations?

* a. ventricles
b. filum terminale
c. cerebellum
d. cerebrum

The cerebellum is located near the occiput and is generally for motor control of the body. The cerebrum is responsible for a wide range of functions which includes: thinking, creativity, emotion, speech, comprehension, figuring out problems, etc.

6. Which cranial nerve of the body is responsible for moving the eye down and in?

a. trigeminal
b. oculomotor
c. optic
* d. trochlear

7. Which cranial nerve of the body functions for hearing?

a. cranial nerve 1
b. cranial nerve 5
c. cranial nerve 6
* d. cranial nerve 8

Cranial nerve 1 (olfactory nerve) functions for our sense of smell. This is a sensory nerve. Cranial nerve 5 (trigeminal nerve) is a mixed nerve. The motor portion is for the function of chewing. The sensory portion is for sensations of pain and touch in specific regions of the face. The three branches of this nerve are the mandibular, maxillary and ophthalmic portions. Cranial nerve 6 (abducens nerve) innervates the lateral rectus muscle of the eyeball. Cranial nerve 8 (vestibulocochlear nerve) is a sensory nerve.

8. Which portion of the brainstem is located between the medulla and midbrain?

a. thalamus
* b. pons
c. cerebrum
d. cerebellum

9. The outer region of the cerebrum is referred to as the ________ and is

made of gray matter.

a. tectum
b. cerebellum
* c. cerebral cortex
d. occipital lobe

The tectum is the posterior region of the midbrain. The tectum houses two inferior colliculi and two superior colliculi. The cerebellum is located inferior to the occipital lobe of the cerebral hemisphere.

10. Which of the following divides the left and right hemispheres of the cerebrum?

* a. longitudinal fissure
b. cerebellum
c. central sulcus
d. transverse fissure

11. The parietal and frontal lobes are divided by which of the following?

a. transverse fissure
* b. central sulcus
c. longitudinal fissure
d. lateral central sulcus

The cerebellum and the cerebrum are divided by the transverse fissure. The lateral central sulcus divides the temporal lobe from the frontal lobe.

12. The folds of the brain are referred to as the __________. Hence, there are ________ lobes in the entire brain.

a. gyri, 4
* b. gyri, 8
c. cerebri, 4
d. cerebri, 8

13. Most of the diencephalon is composed of which of the following?

a. hypothalamus
b. pons
c. cerebellum
* d. thalamus

The thalamus aids in feeling pain, pressure and different temperatures. It forms some of the walls of the third ventricle. The hypothalamus is inferior

to the thalamus. The thalamus helps to send messages to the cerebral cortex. Important thalamic nuclei include lateral geniculate, medial geniculate and ventral posterior. They are responsible for the sense of pain, pressure, touch and vibration. In other words, it receives all information from our senses except for the sense of smell.

14. Which of the following is an anterior structure that has motor tracts located on the medulla?
* a. pyramids
  b. inferior colliculi
  c. hypothalamus
  d. superior colliculi

15. Which is a portion of the brainstem that is anterior to the cerebellum and contains nuclei for cranial nerves 9, 10, 11 and 12?
  a. thalamus
  b. midbrain
  c. pons
* d. medulla oblongata

The pons contains nuclei for cranial nerves 5, 6, 7 and 8. Cranial nerve 8 actually begins from the junction between the medulla oblongata and the pons. The midbrain is also called the mesencephalon. The anterior region of the midbrain has two structures referred to as the cerebral peduncles. The cerebral peduncles contain sensory and motor fibers. The midbrain also contains the substantia nigra, which helps in the control of muscles and attention.

16. Which of the following divides the cerebellum from the cerebrum?
  a. longitudinal fissure
* b. transverse fissure
  c. precentral gyrus
  d. central sulcus

17. All of the following are functions of the hypothalamus except:
  a. temperature regulation
  b. feeling of hunger
  c. feeling of thirst
* d. facial expression

The hypothalamus also functions in our feelings of fear and contentment; it is made up of many nuclei. There are also specific regions of the hypothalamus which are the peptic, supraoptic, tuberal and mammillary. The hypothalamus is the smaller section of the diencephalon. Facial expression is a function of the facial nerve, which is cranial nerve number 7. Smiling, frowning and whistling are all motor functions of the facial nerve.

18. Which of the following is a region of the hypothalamus that aids in giving the brain messages for the sensation of smell?

a. supraoptic
b. inferior colliculus
c. red nucleus
* d. mammillary

19. The most anterior lobe of the __________ is identified as the __________.

* a. cerebrum, frontal lobe
b. cerebrum, parietal lobe
c. cerebellum, temporal lobe
d. cerebellum, parietal lobe

The most posterior lobe of the cerebrum is the occipital lobe. The parietal lobes are anterior and somewhat superior to the occipital lobe. The frontal lobe is anterior to the parietal lobes. The temporal lobes are anterior to the occipital lobe and inferior to the parietal lobes. There are two parietal lobes, two temporal lobes, one frontal lobe and one occipital lobe.

20. Which cranial nerve has a motor portion that innervates the trapezius and sternocleidomastoid muscles?

a. facial nerve
b. abducens nerve
c. vagus nerve
* d. accessory nerve

21. Which of the following cranial nerves is responsible for the taste sensation of sweet?

* a. cranial nerve 7
b. cranial nerve 9
c. cranial nerve 11

d. cranial nerve 12

Cranial nerve 7 (facial nerve) is responsible for the tastes of sweet, sour and salt. Cranial nerve 12 (hypoglossal nerve) is responsible for motor movement of the tongue as in swallowing and speaking. Cranial nerves 5, 7, 9 and 10 are generally considered mixed nerves. This means they have both have a major motor and sensory portion.

22. Cerebral spinal fluid is produced in which of the following locations?

a. cerebellum
* b. choroid plexuses
c. cauda equina
d. brachial plexus

23. Which cranial nerve has a motor portion that functions for smooth muscle contraction for the pharynx and larynx?

a. trochlear
b. abducens
c. facial
* d. vagus

The trochlear nerve (cranial nerve IV) originates from the midbrain. It innervates the superior oblique muscle of the eye. The abducens nerve (cranial nerve VI) originates from the pons. It innervates the lateral rectus muscle of the eye. The facial nerve (cranial nerve VII) starts off from the pons. The anterior two-thirds of the tongue for taste is one function of the facial nerve. The vagus nerve (cranial nerve X) starts off from the medulla. It also gives the sensation of taste from the epiglottis. In addition, it innervates the heart and smooth muscles of the digestive tract.

24. Which cranial nerve is responsible for the sensation of bitter at the back of the tongue?

a. cranial nerve V
b. cranial nerve VI
* c. cranial nerve IX
d. cranial nerve X

25. Which of the following is not a structure of the medulla?

a. olive
b. decussation of pyramids

*   c. tectum
    d. home of cranial nerve nuclei 8-12

The decussation of pyramids is the area where the pyramids cross; they aid in relaying messages relating to muscle movement. On each side of the medulla is a structure referred to as the olive. The tectum is a structure found on the midbrain.

26. Which cranial nerve innervates the lateral rectus muscle which is responsible for moving the eye laterally?
    a. cranial nerve II
    b. cranial nerve III
    c. cranial nerve IV
*   d. cranial nerve VI

27. The center portion of the cerebellum is most closely identified as the:
*   a. vermis
    b. transverse fissure
    c. tentorium cerebelli
    d. longitudinal fissure

The cerebellum is divided into two cerebellar hemispheres. The vermis is between the two cerebellar hemispheres. The transverse fissure divides the cerebellum from the cerebrum. The cerebellum is also divided from the cerebrum by a structure called the tentorium cerebelli, which is made of dura mater.

28. All of the following are functions of the medulla except:
    a. coughing
    b. vomiting
*   c. moving the eyeball down and in
    d. sneezing

29. All of the following are projections into the brain that are made of dura mater except:
    a. falx cerebri
    b. falx cerebelli
    c. tentorium cerebelli
*   d. thalamus

The falx cerebri is located between the cerebral hemispheres. The falx

cerebelli is found between the cerebellar hemispheres. The tentorium cerebelli divides the cerebellar from the occipital lobe of the cerebrum. The lobes of the cerebellum consist of the flocculonodular lobe, anterior lobe and posterior lobe. The flocculonodular lobe has to do with equilibrium. The outer region of the cerebellum is referred to as the cerebral cortex.

30. Which structure attaches the brain stem with the cerebellum?
a. transverse fissure
* b. cerebellar peduncles
c. tentorium cerebelli
d. falx cerebelli

31. Which of the following attaches the cerebellum with the medulla?
a. superior cerebellar peduncles
* b. inferior cerebellar peduncles
c. middle cerebellar peduncles
d. folia

Folia are ridges located on the outer portion of the cerebellum. The superior cerebellar peduncles attach the midbrain with the cerebellum. The middle cerebellar peduncles attach the pons with the cerebellum.

32. The superior peduncles attach which of the following with the cerebellum?
a. pons
b. medulla
* c. midbrain
d. spinal cord

33. Which of the following has just sensory (afferent) fibers that bring messages from the middle portion of the brainstem to the cerebellum?
a. superior cerebellar peduncles
b. anterior lobe
c. inferior cerebellar peduncles
* d. middle cerebellar peduncles

The middle portion of the brainstem is called the pons. The inferior cerebellar peduncles have both sensory (afferent) and motor (efferent) fibers that bring messages from the medulla to the cerebellum and from the cerebellum to the medulla. The superior cerebellar peduncles have both

sensory (afferent) and motor (efferent) fibers. However, the majority is predominately motor fibers.

34. The white matter tracts found deep in the cerebellum are identified as which of the following?

a. falx cerebelli
* b. arbor vitae
c. cerebellar nuclei
d. cerebellar cortex

35. The glossopharyngeal nerve (cranial nerve IX) begins in the __________. It has a function for the taste of ___________ in the posterior 1/3 of tongue.

a. pons, sweet
b. pons, bitter
* c. medulla, bitter
d. midbrain, salt

The facial nerve (cranial nerve VII) is responsible for the taste of sweet, salt and sour in the anterior 2/3rds of the tongue. This nerve also functions in secretion of tears and saliva. Cranial nerve IX functions in our ability to swallow and in the secretion of the salivary glands.

36. Which of the following is one function of the trigeminal nerve (cranial nerve V)?

a. speech
* b. mastication
c. visual acuity
d. lateral movement of the eye

37. Which area of the cerebrum is associated with the function of putting thoughts and words into speech?

a. parietal lobe
b. occipital lobe
* c. Broca area
d. limbic system

The parietal lobe of the cerebrum is associated with analyzing and figuring out certain sensations like temperature, texture, shape, size, two-point discrimination and pain. The occipital lobe helps in figuring out and

understanding visual input. The Broca area is located in the frontal lobe usually in the left hemisphere. The limbic system is located at the interior border of the cerebrum. The limbic system is a very important aspect of the brain that controls our behavior and emotions. It is responsible for our emotions of anger, affection, love, fear, and peacefulness to name a few. Some parts of the limbic system include the dentate gyrus, cingulated gyri, anterior nucleus of thalamus and olfactory bulbs.

38. Which of the following is not part of the limbic system?
a. olfactory bulbs
b. septal nuclei
c. dentate gyrus
* d. cerebellum

39. Which is a portion of the brain that is associated with reading comprehension and understanding spoken words?
a. limbic system
* b. temporal lobe
c. cerebellum
d. pons

The exact area in the temporal lobe that is associated with reading comprehension and understanding spoken words is Wernicke's speech area. The temporal lobe also is associated with one's personality and how one acts around others.

40. Cerebrospinal fluid (CSF) flows into the fourth ventricle from the third ventricle by way of the:
* a. aqueduct of Sylvius
b. foramina of Monroe
c. median aperture of Magendie
d. lateral aperture

41. Cerebrospinal fluid (CSF) enters the third ventricle from the lateral ventricles by which of the following structures?
a. lateral aperture
b. median aperture of Magendie
c. aqueduct of Sylvius
* d. foramen of Monroe

The two lateral apertures and one median aperture of Magendie are located in the fourth ventricle. These structures are the passageways where CSF can enter the subarachnoid space.

42. The reabsorption of CSF in the body takes place in which location?
    a. pia mater
*   b. arachnoid villi
    c. central sulcus
    d. cerebral cortex

43. All of the following are structures of the cerebrum except:
    a. postcentral gyrus
    b. precentral gyrus
    c. longitudinal fissure
*   d. pons

Other structures of the cerebrum include the central sulcus, cerebral cortex, cerebral lobes, basal ganglia and corpus striatum. The basal ganglia are large groups of nuclei located in the cerebrum; it is related with motor and learning functions. The corpus striatum contains the caudate nucleus, lentiform nucleus, globus pallidus and putamen.

44. The majority of the diencephalon is composed of which structure that is not part of the cerebellum?
*   a. thalamus
    b. vermis
    c. hypothalamus
    d. tentorium cerebelli

45. Which of the following has the lenticular nucleus, putamen and caudate nucleus?
    a. pituitary gland
    b. cerebellum
*   c. corpus striatum
    d. caudate nucleus

The corpus striatum is located in both hemispheres of the brain. The globus pallidus is medial to the putamen, which both make up the lenticular nucleus. The caudate nucleus is superior to the globus pallidus.

46. All of the following are parts of the basal ganglia except:

a. lenticular nucleus
b. corpus striatum
c. globus pallidus
* d. falx cerebelli

47. Which portion of the cerebrum is located under the frontal, parietal and temporal cerebral lobes?

* a. insula
b. cerebral cortex
c. cerebellum
d. precentral gyrus

The precentral gyrus is anterior to the postcentral gyrus. The frontal and temporal lobes are divided from each other by way of the lateral cerebral sulcus. The cerebellum contains anterior, posterior and flocculonodular lobes.

48. All of the following cranial nerves are associated with the eye except:

a. cranial nerve II
b. cranial nerve III
* c. cranial nerve XI
d. cranial nerve VI

49. Which structure of the brain is significantly associated with the movement of muscles in the body?

a. hypothalamus
b. medulla
* c. cerebellum
d. pons

The hypothalamus helps with feelings of when we are hungry and thirsty. Also, it helps us in sustaining our sleep. The medulla is the most inferior portion of the brain stem. It helps sustain breathing and in maintaining and managing the heart rate. The pons is the middle portion of the brainstem. It also helps maintain breathing. Most of the cranial nerves begin at the medulla and pons.

50. The main motor area is found in which portion of the cerebral cortex?

* a. precentral gyrus of frontal lobe

b. postcentral gyrus
c. temporal lobe
d. occipital lobe

51. The primary olfactory area of the brain is found in which of the following regions?
a. occipital lobe
* b. temporal lobe
c. precentral gyrus of frontal lobe
d. postcentral gyrus

The olfactory area is responsible for the sense of smell. The postcentral gyrus houses the primary gustatory area, which is responsible for the sense of taste. The occipital lobe houses the primary visual area.

52. The primary gustatory area is responsible for the sense of _________. This area is found in the area of the _________.
a. sound, temporal lobe
b. sight, occipital lobe
* c. taste, postcentral gyrus
d. smell, temporal lobe

53. Which area of the cerebrum is the primary auditory area situated?
a. parietal lobe
b. frontal lobe
c. occipital lobe
* d. temporal lobe

The primary visual area, primary auditory area, primary olfactory area and primary gustatory area are all located in the cerebrum. The primary visual area is located in the occipital lobe of the cerebrum.

54. Which of the following is identified as Wernicke's area of the brain?
a. visual association area
b. gnostic area
* c. auditory association area
d. primary olfactory area

55. The interpretation of different sensations based on past sensory memories is a function of which association area?

a. frontal eye field area
b. primary olfactory area
* c. somatosensory association area
d. premotor area

The primary olfactory area can be eliminated as a choice immediately since it is a sensory area. The premotor area is a type of an association area that deals with very complex motor movements. For example, the hand and finger movement used to paint a picture is a function of the premotor area. The frontal eye field area is a type of an association area that is responsible for scanning by the eyes. For example, this part of the brain might be highly developed in athletes who play baseball or tennis. This is because these types of sports require fast scanning capability of the eyes. In baseball, having a fast scanning capability can give a fielder a faster reaction time to a ball that is hit.

56. Which two of the following are considered motor areas of the cerebrum?
* a. Boca's area, primary motor area
b. somatosensory association, primary olfactory area
c. primary auditory area, primary motor area
d. gnositc area, premotor area

57. Which functional area is a type of sensory area in the cerebrum?
* a. primary somatosensory area
b. visual association area
c. primary motor area
d. premotor area

The primary somatosensory area is responsible for the sense of touch. It also aids in feeling pain, temperature and in knowing the positioning of one's joints.

58. Which of the following are considered sensory areas of the cerebrum?
a. primary visual area, auditory association area
* b. primary somatosensory area, primary visual area
c. auditory association area, primary motor area
d. primary visual area, visual association area

59. Which of the following is not a structure of the brain stem?

a. medulla
b. pons
* c. cerebellum
d. midbrain

The medulla is the most inferior portion of the brain stem. The pons is superior to the medulla. The midbrain is the most superior portion of the brain stem. The cerebellum is posterior to the brain stem.

60. Which cranial nerve is responsible for our sense of smell?
a. trochlear nerve
b. trigeminal nerve
* c. olfactory nerve
d. vagus nerve

61. Which of the following has nuclei located in the pons?
* a. trigeminal nerve
b. oculomotor nerve
c. trochlear nerve
d. vagus nerve

The nuclei of the abducens nerve, facial nerve and the vestibulocochlear nerve are also found in the pons. The nuclei of the oculomotor and trochlear nerves are found in the midbrain. The nuclei of the vagus nerve are located in the medulla.

62. Which of the following is not a structure of the midbrain?
a. inferior follicular
b. tectum
c. corpora quadrigemina
* d. olive

63. Which cranial nerve has mandibular, maxillary and ophthalmic portions?
a. trochlear nerve
* b. trigeminal nerve
c. olfactory nerve
d. vagus nerve

Actions of chewing and sensations of pain and touch throughout the face are made possible by the trigeminal nerve. In addition, sensations of pain and

touch at inside portions of the mouth are made possible by cranial nerve five.

Matching

64. smiling, frowning and raising of the eyes are functions of this cranial nerve	a. vestibulocochlear nerve
65. hearing and equilibrium	b. optic nerve
66. innervates the levator palpebrae	c. facial nerve
67. innervates most extrinsic eyeball muscles, which are the inferior oblique, superior rectus, medial rectus and inferior rectus	d. oculomotor nerve
68. taste of bitter on the posterior 1/3 of tongue is made possible by this cranial nerve	e. abducens nerve
69. this cranial nerve innervates the lateral rectus muscle which moves the eyeball laterally	ab. glossopharyngeal nerve
70. this cranial nerve innervates the trapezius and the sternocleidomastoid	ac. hypoglossal nerve
71. this cranial nerve is vital for tongue movements to aid in speech	ad. accessory nerve

Answers

64. c 65. a 66. d 67. d 68. ab 69. e

70. ad 71. ac

Neural Integrative, Motor & Sensory Systems

1. Which type of sensory receptor is specifically responsible for sensing pain?

a. thermoreceptor

b. chemoreceptor

* c. nociceptor

d. mechanoreceptor

Thermoreceptors are responsible for sensing temperature. Chemoreceptors are responsible for sensing chemicals by smell and taste. Mechanoreceptors are responsible for sensing pressure.

2. Which of the following is not linked with simple receptors?

a. pain

b. pressure

* c. vision

d. proprioception

3. Which of the following is responsible for the sense of position of joints and muscles?

* a. proprioceptors

b. thermoreceptors

c. photoreceptors

d. mechanoreceptors

Photoreceptors are responsible for sensing light. Proprioception is the sense of knowing where body parts are located in relation to other body parts. For example, knowing the difference of where your foot is in relation to your hand is having a sense of proprioception. Golgi tendon organs are types of proprioceptors.

4. Which of the following are sites for proprioceptors?

a. tendons and joints

b. joints and muscles

c. ear

* d. all of the above

5. Senses such as touch, smell, hearing, equilibrium and pain are made possible by which of the following that gives us information about our

surroundings?

* a. exteroceptors
b. proprioceptors
c. thermoreceptors
d. stereognosis

Exteroreceptors are accountable for pressure, touch, hearing, smell, taste, sight, pain, temperature and vibration. Stereognosis is the ability to identify an object by utilizing the sense of touch.

6. Which of the following is crucial for the transmission of nerve impulses for proprioception that is subconscious?

a. lateral spinothalmic tract
b. corticobulbar tract
* c. anterior spinocerebellar tract
d. anterior corticospinal tract

7. Which sensory pathway transmits crude touch, temperature, pain, pressure and itch from the left or right side of the body to the right or left cerebral hemisphere?

a. posterior spinocerebellar tract
* b. spinothalamic tract
c. posterior column-medial lemniscus
d. tectospinal tract

The spinothalamic pathway is composed of the lateral spinothalamic tract and the anterior spinothalamic tract. The lateral spinothalamic tract is responsible for sensations of different temperatures and levels of pain. The anterior spinothalamic tract is in charge of sensations of crude touch, pressure and itch. The posterior column-medial lemniscus pathway is responsible for discriminative touch. Discriminative touch is a sensation that is quite extensive in the fingers. The posterior spinocerebellar tract also is in charge of proprioception. The tectospinal tract conveys impulses that are motor for movements of the head and eyes.

8. All of the following give impulses that are motor to one side of the brain to the contrary or opposite section of the body except:

a. tectospinal tract
b. rubrospinal tract
* c. lateral spinothalmic tract

d. lateral corticospinal tract

9. All of the following give impulses that are sensory to one side of the body to the contrary or opposite side of the brain except:
a. lateral spinothalamic tract
b. anterior spinothalamic tract
* c. rubrospinal tract
d. posterior column-medial lemniscus

The lateral spinothalamic tract and anterior spinothalamic tract are sensory tracts. The posterior column-medial lemniscus pathway is a sensory pathway. In other words, this is an ascending pathway. The motor pathways are divided into pyramidal and extrapyramidal tracts. The corticobulbar tract, lateral corticospinal and anterior corticospinal tract are all pyramidal tracts. The tectospinal tract, vestibulospinal tract and the rubrospinal tract are the primary extrapyramidal tracts.

10. ___________ is (are) not a function of the posterior column-medial lemniscus pathway, which is a sensory pathway.
a. Weight discrimination
b. Stereognosis
c. Proprioception
* d. Feelings of affection

11. Which of the following is considered a special sense?
* a. vision
b. pain
c. proprioception
d. vibration

Hearing, equilibrium, taste, vision and smell are special senses. Pain, proprioception, vibration, temperature sensation, and pressure are general senses. Complex receptors are related with special senses, and simple receptors are linked with general senses. Remember that special senses are located in the head.

12. All of the following are considered cutaneous sensations except:
a. vibration
b. feeling of heat
c. pain

* d. equilibrium

13. Which of the following is considered a sensory tract?
    a. tectospinal tract
    b. vestibulospinal tract
*   c. anterior spinocerebellar tract
    d. anterior corticospinal tract

The vestibulospinal tract is a motor tract that has to do with balance and equilibrium. The anterior corticospinal tract is a motor tract that is responsible for skeletal muscle movement.

14. Which of the following is an extrapyramidal tract that has to do with movement of the legs and arms?
    a. vestibulospinal tract
*   b. rubrospinal tract
    c. corticobulbar tract
    d. posterior spinocerebellar tract

15. Which of the following is not associated with a pyramidal tract?
    a. motor tract
    b. descending tract
    c. direct tract
*   d. sensory tract

Ascending tracts are associated with sensory tracts. Indirect tracts are also known as extrapyramidal tracts. Pyramidal and extrapyramidal tracts are motor tracts.

16. The tectospinal tract begins at which location?
    a. cerebrum
    b. pons
*   c. midbrain
    d. medulla

17. All of the following are extrapyramidal tracts except:
    a. rubrospinal tract
*   b. lateral corticospinal tract
    c. vestibulospinal tract
    d. tectospinal tract

The lateral corticospinal tract is a pyramidal tract, which is motor. The vestibulospinal tract is concerned with balance, and the rubrospinal tract is responsible for muscle movement of the arms and legs.

18. The axons of first order neurons are located in which area of the posterior column-medial lemniscus pathway?

* a. fasciculus gracilis
b. cerebellum
c. caudia equina
d. nucleus cuneatus

19. Which of the following has axons that enter the medial lemnicus in the posterior column-medial lemniscus pathway by going to the opposite side of the medulla?

a. first-order neurons
* b. second-order neurons
c. third-order neurons
d. cerebrum

First-order neurons have axons that are located in the fasciculus cuneatus and fasciculus gracilis of the posterior white column. The end of these first-order neurons connect with the second-order neurons at the lowest portion of the brainstem. This portion is at the medulla. The second-order neurons travel to the opposite side of the medulla and then superior to the medial lemniscus in the midbrain and finally to the thalamus. The second-order neurons meet up with the third-order neurons in the thalamus. The third-order neurons then connect to the somatosensory region of the cerebral cortex.

20. Which is the correct order of nerve impulse conduction of the posterior column-medial lemniscus pathway?

* a. fasciculus cuneatus, nucleus cuneatus, opposite side of medulla, medial lemniscus, thalamus, cerebral cortex
b. nucleus cuneatus, posterior white column, opposite side of medulla, medial lemniscus, cerebral cortex, thalamus
c. cerebral cortex, nucleus cuneatus, posterior white column, thalamus, opposite side of medulla, fasciculus cuneatus
d. fasciculus cuneatus, thalamus, medial lemniscus, opposite side of medulla, thalamus, cerebral cortex

21. Third order neurons of the posterior column-medial lemniscus pathway are located in the:

a. fasciculus cuneatus
b. midbrain
c. medulla
* d. cerebral cortex

The posterior column-medial lemniscus pathway is an ascending pathway. This means that it is a sensory pathway. This pathway receives information from sensory receptors. The correct message path from first to last is the following: spinal cord, medulla, midbrain, thalamus, cerebral cortex.

22. Which of the following is a function of the posterior column-medial lemniscus pathway?

a. sight
b. sound
* c. discriminative touch
d. smell

23. Golgi tendon organs are associated with which of the following?

a. hearing
b. taste
* c. proprioception
d. emotion

Golgi tendon organs can figure out and monitor how much a muscle is being contracted. If the muscle is being contracted too much, they send messages to the muscle to stop this action. This is a protection mechanism to prevent muscle and tendon injury.

24. Control of muscles in the cervical region is most closely associated with which motor tract?

* a. tectospinal tract
b. anterior spinothalamic tract
c. lateral corticospinal tract
d. lateral spinothalamic tract

Autonomic Nervous System

1. Which of the following is not controlled by the autonomic nervous system?

a. smooth muscle
b. cardiac muscle
* c. skeletal muscle
d. specific glands

The autonomic nervous system is composed of the parasympathetic and sympathetic divisions. The autonomic nervous system is not usually under voluntary control.

2. Which of the following is not an effect of the parasympathetic division?

* a. increase in heart rate
b. intestine relaxes
c. blood pressure decreases
d. constriction of pupils

3. Which of the following is part of the peripheral nervous system?

a. somatic nervous system
b. parasympathetic division
c. sympathetic division
* d. all of the above

The sympathetic division of the autonomic nervous system is the flight-or-flight division. Some effects of this division include an increase in heart rate, increase in blood pressure, dilation or enlargement of the pupils of the eyes, and the break down of fat in the body. Some effects of the parasympathetic division of the autonomic nervous system include a decrease in heart rate, increase in saliva, constriction of pupils and a decrease in blood pressure.

4. All of the following are effects of the sympathetic nervous system except:

* a. decrease in saliva
b. heart rate goes up
c. blood pressure increases
d. increase in sweat

5. Which of the following is an autonomic motor neuron that has a myelinated axon?

a. cell body
b. postganglionic fiber
* c. preganglionic neuron
d. postganglionic neuron

A preganglionic fiber is also known as a myelinated axon. Postganglionic neurons are completely located in the PNS. The cell body of a preganglionic neuron is located in the CNS.

6. The somatic nervous system regulates which of the following?

a. glands
b. smooth muscle
c. cardiac muscle
* d. skeletal muscle

7. Which of the following is myelinated and a portion is located in the CNS?

* a. preganglionic neurons
b. postganglionic neurons
c. postganglionic fiber
d. spinal nerve

A spinal nerve is located in the PNS. A postganglionic fiber is an unmyelinated axon that is part of the postganglionic neuron.

8. Which of the following is not a characteristic of a postganglionic fiber?

a. It is an axon.
* b. It is myelinated.
c. It is unmyelinated.
d. It is located in the PNS.

9. All of the following are locations of the cell bodies of preganglionic neurons for the parasympathetic division except:

a. nuclei of cranial nerve 3
b. nuclei of cranial nerve 7
c. nuclei of cranial nerve 9
* d. cell body in lateral gray horn of T10

The nuclei of cranial nerve 10 and the lateral gray horns of segments S2-S4

are the other locations of the cell bodies of preganglionic neurons for the parasympathetic division. The cell bodies of preganglionic neurons in the sympathetic division are located in the lateral gray horns from segments T1-L2. The sympathetic division is also referred to as the thoracolumbar division, and the parasympathetic division is also identified as the craniosacral division.

10. Which of the following is part of the sympathetic division of the autonomic nervous system?

a. S4 sacral nerve
b. S3 sacral nerve
* c. T2 spinal nerve
d. nuclei of cranial nerve III

11. All of the following are associated with the sympathetic division of the autonomic nervous system except:

a. fight or flight
b. thoracolumbar division
c. T5 spinal nerve
* d. rest and digest

Resting and digesting are associated with the parasympathetic division. The craniosacral division entails the oculomotor nerve, facial nerve, glossopharyngeal nerve, and vagus nerve. Sacral nerves S2, S3 and S4 are also included in this division.

12. Preganglionic axons of the sympathetic division depart the spinal cord in which of the following locations?

* a. anterior root of spinal nerve
b. posterior root of spinal nerve
c. sympathetic trunk ganglion
d. posterior white column

13. Which of the following links the anterior ramus of a spinal nerve with the sympathetic trunk ganglion?

a. anterior root of a spinal nerve
b. posterior root ganglion
* c. white ramus communicans
d. posterior root of a spinal nerve

The anterior root of a spinal nerve is connected directly to the front of the spinal cord. The posterior root of a spinal nerve is connected directly to the back of the spinal cord.

14. Which of the following innervates the heart and is located near C7-T1?
  a. superior cervical ganglion
  b. greater splanchnic nerve
  c. lesser splanchnic nerve
* d. inferior cervical ganglion

15. The middle cervical ganglion and the ____________ both innervate the heart. They are part of the sympathetic division of the autonomic nervous system.
* a. inferior cervical ganglion
  b. superior cervical ganglion
  c. oculomotor nerve
  d. facial nerve

The superior cervical ganglion, middle cervical ganglion and inferior cervical ganglion make up the cervical section of the sympathetic trunk. The superior cervical ganglion is located near the upper cervical area at approximate level of C2. The middle cervical ganglion is found at the level of the C5-C6 region. The lower cervical ganglion is located at the level of the C7-T1 region.

16. Which of the following receives preganglionic fibers from the oculomotor nerve in the parasympathetic division?
  a. myenteric plexus
  b. submandibular ganglion
  c. otic ganglion
* d. ciliary ganglion

17. Postganglionic fibers are extended to the lacrimal glands from the _________ of the parasympathetic division of the autonomic nervous system.
  a. ciliary ganglion
* b. pterygopalatine ganglion
  c. myenteric plexus
  d. otic gangion

The oculomotor nerve sends preganglionic fibers to the ciliary ganglion.

Postganglionic fibers are extended to the ciliary eye muscles and other smooth muscles by way of the ciliary ganglion. The vagus nerve connects to the myeteric and Auerbach plexuses by way of the preganglionic fibers. Postganglionic fibers are extended from the myenteric and Auerbach plexuses to the glands and smooth muscle of organs in the thoracoabdominal cavity. A plexus is a linked group of ganglia. The glossopharyngeal nerve connects to the otic ganglion by way of preganglionic fibers. Postganglionic fibers are extended from the otic ganglion to the parotid gland.

18. Which of the following connects to the otic ganglion by way of preganglionic fibers?

a. facial nerve
b. vagus nerve
* c. glossopharyngeal nerve
d. oculomotor nerve

19. All of the following are locations where postganglionic fibers are extended from the pterygopalatine ganglion except:

a. lacrimal gland
b. palate
* c. parotid gland
d. pharynx

The parotid gland has postganglionic fibers that extend to it by way of the otic ganglion. The glossopharyngeal nerve connects to the pterygopalatine ganglion by way of the preganglionic fibers.

20. Pelvic splanchnic nerves synapse with which of the following?

* a. parasympathetic postganglionic neurons
b. ciliary ganglion
c. submandibular ganglion
d. otic ganglion

21. Which of the following nerves is associated with the sympathetic division of the autonomic nervous system?

* a. lumbar splanchnic nerve
b. pelvic splanchnic nerve
c. facial nerve

d. glossopharyngeal nerve

The pelvic spanchnic nerves, facial nerve and glossopharyngeal nerve are associated with the parasympathetic division of the autonomic nervous system. The vagus nerve, oculomotor nerve and sacral nerves S2-S4 are also associated with the parasympathetic division.

22. The sigmoid colon, rectum, kidney and sex organs are all innervated by the:

a. glossopharyngeal nerve
b. facial nerve
* c. pelvic splanchnic nerves
d. oculomotor nerve

23. Which of the following is a parasympathetic division function?

* a. secretion of lacrimal gland
b. dilation of pupil
c. relaxation of ciliary eye muscle
d. increase in sweat secretion in feet and hands

Contraction of ciliary eye muscle, salivary gland secretion and constriction of the pupil are parasympathetic division functions. Relaxation of ciliary eye muscle, increase in sweat secretion, and dilation of the pupil are all sympathetic division functions.

24. Which of the following neurotransmitters are released by adrenergic neurons in the autonomic nervous system?

a. acetylcholine, epinephrine
b. acetylcholine
* c. epinephrine, norepinephrine
d. norepinephrine, acetylcholine

25. Which of the following are *usually* classified as adrenergic?

a. sympathetic preganglionic neurons
* b. sympathetic postganglionic neurons
c. parasympathetic preganglionic neurons
d. parasympathetic postganglionic neurons

Sympathetic preganglionic neurons, parasympathetic preganglionic neurons, parasympathetic postganglionic neurons are classified as cholinergic. There is a very tiny portion of sympathetic postganglionic neurons that are also

classified as cholinergic. However, for practical purposes the majority of sympathetic postganglionic neurons are adrenergic.

26. Which of the following is a neurotransmitter that is released by cholinergic neurons?

a. epinephrine
b. norepinephrine
* c. acetylcholine
d. adrenalin

27. Motor messages or impulses are sent from the spinal cord or brain to the ganglion by which of the following?

a. postganglionic neurons
* b. preganglionic neurons
c. effectors
d. sensory neuron

Motor messages or impulses are sent from the automatic ganglion to either glands, smooth muscles or cardiac muscles. Glands, smooth muscles and cardiac muscles are the effectors.

28. All of the following are effectors except:

a. glands
b. smooth muscle
* c. ganglion
d. cardiac muscle

29. Nicotinic receptors are usually located on parasympathetic postganglionic neurons and which of the following?

* a. sympathetic postganglionic neurons
b. effectors that are smooth muscles
c. effectors that are glands
d. effectors that are cardiac muscles

Effectors, which are smooth muscle, glands and cardiac muscle, have muscarinic receptors. Muscarinic receptors and nicotinic receptors are kinds of cholinergic receptors.

30. Which of the following is a sympathetic response to a stimulus?

a. heart rate decreases

*   b. dilation of blood vessels of skeletal muscle
          c. decrease in blood pressure
          d. insulin secretion increases

31. Which of the following are adrenergic receptors?
*   a. alpha receptor, beta receptor
          b. nicotinic receptor, beta receptor
          c. nicotinic receptor, muscarinic receptor
          d. alpha receptor, muscarinic receptor
Alpha and beta receptors are found on effectors such as smooth muscles, cardiac muscles and glands. Nicotinic receptors and muscarinic receptors are cholinergic receptors.

32. All of the following are most closely associated with cholinergic receptors except:
          a. acetylcholine
          b. muscarinic
          c. nicotinic
*   d. epinephrine

33. Urination and defacation are considered which types of responses?
          a. fight-or-flight responses
          b. sympathetic responses
*   c. parasympathetic responses
          d. defense responses
Some other parasympathetic responses are decreases in blood pressure, decreases in heart rate, increases in insulin levels, contraction of the gallbladder and narrowing of heart vessels.

34. ___________ are types of autonomic ganglia that are sympathetic and located __________ to the spine.
          a. Paravertebral ganglia, anterior
*   b. Prevertebral ganglia, anterior
          c. Prevertebral ganglia, posterior
          d. Paravertebral ganglia, posterior
Paravertebral ganglia are also sympathetic and located on each side of the spine. Paraveretebral ganglia also include the superior mesenteric ganglia, inferior mesenteric ganglia and celiac ganglia.

35. Which of the following is least closely associated with adrenergic receptors?

a. beta
b. norepinephrine
* c. acetylcholine (ACh)
d. alpha

36. Which of the following is not a parasympathetic function?

a. reduce heart rate
b. erection of sexual organs
* c. skeletal muscle contraction
d. increase heart rate

37. Which of the following nerves connects to the inferior mesenteric ganglion?

* a. L1-L3 splanchnic nerve
b. T12 least splanchnic nerve
c. T5-T9 greater splanchnic nerve
d. T10-T11 lesser splanchnic nerve

T12 least splanchnic nerve and T10-T11 lesser splanchnic nerve connect to the superior mesenteric ganglion. T5-T9 greater splanchnic nerve connects to the celiac ganglion.

38. In the sympathetic division, postganglionic fibers of which ganglion innervate the ascending colon?

* a. superior mesenteric
b. ciliary ganglion
c. otic ganglion
d. submandibular ganglion

39. Preganglionic fibers of the L1-L3 lumbar splanchnic nerve synapses with postganglionic fibers in which of the following?

a. celiac ganglion
* b. inferior mesenteric ganglion
c. superior mesenteric ganglion
d. otic ganglion

The postganglionic fibers related to the L1-L3 lumbar splanchnic nerve innervate the descending colon, rectum, urinary bladder and sex organs.

Special Senses

1. All of the following are structures of the middle ear except:
   a. incus
   b. stapes
   c. malleus
* d. vestibule

The vestibule is a structure of the inner ear. The vestibule is the central region of the bony labyrinth. The vestibule contains the saccule and utricle. Some other structures of the middle ear are the tympanic cavity, tympanic antrum and oval window. The stapes attaches to the oval window.

2. Which of the following is not a considered an auditory ossicle?
   a. malleus
   b. stapes
* c. tympanic antrum
   d. incus

3. The outer visible region of the ear that is functional for gathering sound waves is called the:
   a. stapes
* b. auricle
   c. saccule
   d. malleus

The outer portion of the ear includes the eardrum (tympanic membrane), helix, external auditory meatus, ceruminous glands and lobule.

4. Which of the following is located on the superior region of the pinna?
   a. lobule
   b. incus
   c. cochlea
* d. helix

5. The lower floppy portion of the ear is best identified as the:
   a. helix
   b. incus
   c. pinna
* d. lobule

The pinna is also known as the auricle, which is the outer visible region of the ear that is exterior to the head.

6. The Eustachian tube is associated with equalizing which of the following?

a. taste
* b. air pressure
c. sight
d. vibration

7. Which of the following structures is referred to as the blind spot?

a. iris
b. lens
c. conjunctiva
* d. optic disc

The lens is found posterior to the iris. It is composed of crystallins. Ciliary muscles are responsible for altering the thickness of the lens. The lens is held in place by suspensory ligaments. This makes it suitable for seeing objects that are near or far away. The conjuctiva is a protective layer that is around the anterior portion of the eye.

8. Which cranial nerve is responsible for holding the eye open?

a. cranial nerve 2
* b. cranial nerve 3
c. cranial nerve 5
d. cranial nerve 6

9. Taste buds are located on papillae. Which types of papillae are found on the posterior area of the tongue?

* a. circumvallate
b. filiform
c. fungiform
d. gustatory hair

Fungiform papillae are located on the anterior tip of the tongue and the lateral sides of the tongue. Most of the anterior 2/3rds of the tongue contains filiform papillae, which contain the least quantity of taste buds.

10. Which of the following are best identified as the eyelids?

a. pinna
b. sclera
c. puncta
* d. palpebrae

11. Which of the following is specifically responsible for our ability to visually see things in color?
a. iris
* b. cones
c. sclera
d. optic disc

The optic disc can be eliminated as a choice right away because it doesn't have rods or cones. The sclera is the white portion of the eye that is seen; it is avascular. The iris contains pigment which gives color to the eye.

12. Which of the following is held in place by suspensory ligaments?
a. iris
b. retina
c. cornea
* d. lens

13. The pupil is a structure that is found in the _________ of the eye; it narrows and dilates.
a. sclera
b. retina
* c. iris
d. lens

The retina is the inner covering of the eye that contains cones and rods. Rods are vital for seeing objects that are black and white. Cones are important for seeing images in color. The retina changes light into nerve impulses. There are blood vessels in the retina that can be observed by a doctor for conditions such as diabetic retinopathy. The lens is made of proteins referred to as crystallins. The pupil is located in the middle layer of the eye.

14. Which of the following is a structure of the retina that is directly responsible for the ability to see objects in dim light?
a. cone

*	b. rod
	c. optic disc
	d. iris

15. Which of the following is not a structure of the retina?
	a. optic disc
	b. superior nasal arteries and veins
	c. macula lutea
*	d. ciliary muscle

The retina is the internal layer of the eye. The ciliary muscle is located in the intermediate layer of the eye. The center area of the back region of the retina is referred to as the macula lutea.

16. Which structure of the anterior eye is normally clear and serves as a source of protection?
	a. sclera
*	b. conjuctiva
	c. iris
	d. retina

17. Which of the following is not part of the intermediate layer of the eye?
*	a. retina
	b. ciliary body
	c. choroid
	d. iris

The retina is part of the inner layer of the eye. The iris is what gives the color to the eye. For example, most people have brown eyes and a smaller percentage green and blue. The optic disc is found in the retina. The optic nerve can be observed by way of the optic disc. The thickest area of the vascular tunic (middle layer of eye) consists of the ciliary body. The ciliary body is made up of the ciliary muscle and ciliary processes.

18. The intermediate layer of the eye that houses the very vascular choroid is called the:
	a. sclera
*	b. uvea
	c. retina
	d. fibrous tunic

19. The anterior chamber of the eyeball is located between the:

* a. cornea and iris
  b. iris to lens
  c. lens and retina
  d. sclera to retina

There are three chambers of fluid in the eye, which are the posterior, anterior and vitreous chambers. The posterior chamber is between the iris and lens. The vitreous chamber is between the lens and retina.

20. Which portion of the eye contains the sclera?

  a. nervous tunic
  b. vascular tunic
* c. fibrous tunic
  d. lens

21. All of the following are structures of the inner ear except:

  a. saccule
  b. utricle
  c. cochlea
* d. incus

The incus is a structure of the middle ear. The incus is considered an auditory ossicle. The incus articulates with the malleus and the stapes. The malleus and stapes are also considered auditory ossicles. The incus, malleus and stapes vibrate whenever the tympanic membrane vibrates in respond to sound waves. The malleus is directly attached to the tympanic membrane. The cochlea is the spiral portion of the inner ear. The utricle and saccule are involved with equilibrium or in the sustaining of balance.

22. The __________ is a structure of the inner ear that contains the organ of Corti, which helps to convey sound impulses to cranial nerve eight.

  a. auricle
  b. tympanic membrane
  c. incus
* d. cochlea

23. Which of the following is a structure that is connected to the pharynx and aids in equalizing pressure between the middle ear and the atmosphere?

* a. Eustachian tube

b. pinna
c. eardrum
d. saccule

Swallowing and yawning can open the Eustachian tube. The Eustachian tube also helps to get rid of mucus from the middle ear. The pinna or the auricle is part of the outer ear. The eardrum or tympanic membrane is part of the outer ear or middle ear depending on the teacher. However, some teachers also consider the tympanic membrane to act as a border of separation between the external ear and middle ear. The saccule is part of the inner ear. The saccule aids in equilibrium.

24. Which of the following equalizes pressure in the middle ear?
a. sneezing
b. yawning
c. swallowing
* d. all of the above

25. All of the following are structures of the inner ear except:
a. bony labyrinth
* b. stapes
c. membranous labyrinth
d. organ of Corti

The stapes is an auditory ossicle and a structure of the middle ear. The organ of Corti is known as the hearing organ. The inner ear contains the membranous labyrinth and bony labyrinth. The semicircular canals in the inner ear aid in balance and in figuring acceleration. There is a posterior semicircular canal, lateral semicircular canal, and an anterior semicircular canal. The membranous labyrinth has endolymph; the bony labyrinth has perilymph.

26. Which two of the following are not technically part of the semicircular canals but part of the vestibule in the membranous labyrinth and also involved with equilibrium?
a. cochlea, organ of Corti
b. malleus, incus
c. incus, stapes
* d. saccule and utricle

27. Which structure of the ear empties into an ampulla?
a. lobule
* b. semicircular canal
c. cochlea
d. eardrum

The lobule is the bottom portion of the auricle. The lobule is a structure of the outer ear. The eardrum is a structure of the external ear. However, some consider it to be part of the middle ear. The cochlea is involved with hearing. The three semicircular canals are involved with equilibrium. Each semicircular canal empties into the ampulla of the semicircular duct.

28. Which of the following is the best example of dynamic equilibrium?
* a. the sensation of acceleration while riding on a roller coaster
b. the loud music at a concert
c. swallowing food
d. the low pitch of a specific musical note

29. Which type of joint is found between the bony ossicles of the middle ear?
a. synchondrosis
b. syndesmosis
* c. synovial
d. synostosis

The tensor tympani is a muscle that is connected to the ossicles. The stapedius is connected to the stapes. These muscles serve to contract the tympanic membrane in response to certain volumes of sound. In this way, sound vibrations are properly transmitted to the rest of the middle and inner ear.

30. All of the following are considered part of the bony labyrinth of the inner ear except:
a. semicircular canal
b. cochlea
c. vestibule
* d. cochlea duct

31. The walls of both the utricle and saccule contain a sensory region called the:

a. tympanic membrane
* b. macula
c. incus
d. helicotrema

The helicotrema is a structure found at the end of the cochlea. The macula is responsible for our sense of static and dynamic equilibrium. The macula has three kinds of cells. There is a type I sensory hair cell, type II sensory hair cell and a supporting or sustentacular cell. The hair cells have one kinocilium and many stereocilia. Between each hair cell are supporting cells which are columnar cells.

32. ________ is a structure of the ________ ear that manufactures the otolithic membrane.
a. Stereocilia, middle
* b. Supporting cell, inner
c. Kinocilium, inner
d. Oval window, middle

33. Which of the following is an example of a photopigment in the rods?
a. ampulla
b. modiolus
* c. rodopsin
d. crystallins

The ampulla and modiolus are structures of the inner ear. Crystallins are proteins found in the lens of the eye.

34. In which location do the optic nerves cross?
* a. optic chiasma
b. optic tract
c. lateral geniculate nuclei
d. optic radiation

35. The surface of the otolithic membrane is occupied by __________, which is (are) found in the __________ ear.
a. malleus, middle
b. incus, middle
* c. otoliths, inner
d. cranial nerve eight, inner

Otoliths are made of calcium carbonate and protein. Otoliths react to gravity and positional changes of the body. In this way, the hair cells are also influenced each time the otoliths and otolithic membrane move. In this way, this can help transmit messages to cranial nerve eight.

36. Which of the following is the most anterior structure?
* a. optic nerve
  b. optic chiasma
  c. lateral geniculate nuclei
  d. optic tract

37. Which structure of the ear is partially found in the oval window?
  a. external auditory meatus
  b. incus
  c. malleus
* d. stapes

The malleus is connected to the tympanic membrane. The head of the malleus is connected to the incus. The incus is connected to the stapes.

38. Which structure of the ear is located most inferior to the oval window?
  a. tympanic antrum
  b. anterior semicircular canal
  c. lateral semicircular canal
* d. round window

39. Which structure of the ear is found closest to the mastoid air cells of the temporal bone?
  a. cochlea
  b. Eustachian tube
* c. tympanic antrum
  d. lobule

The cochlea is a structure of the inner ear. The lobule is a structure of the outer ear. The tympanic antrum and Eustachian tube are part of the middle ear.

40. Which portion of the anterior cavity of the eye is posterior to the cornea, anterior to the iris and helps to sustain the eyeball shape?
* a. anterior chamber

b. posterior chamber
c. retina
d. bulbar conjunctiva

41. The aqueous humor is in the anterior chamber of the eye where it will then go into which of the following?
a. lens
b. retina
c. macula lutea
* d. canal of Schlemm

The canal of Schemm is a venous sinus. This is where the aqueous humor goes to get into the bloodstream. Aqueous humor is secreted by ciliary processes.

42. Which of the following is located between the retina and lens of the eye?
a. anterior cavity
* b. vitreous chamber
c. iris
d. sclera

43. Which of the following is a component of the outer ear?
* a. helix
b. cochlea
c. vestibule
d. semicircular canals

The cochlea, vestibule and semicircular canals are all components of the inner ear. The inner ear contains both perilymph and endolymph. The cochlea aids in conveying sound waves to the organ of Corti.

44. Which two of the following are important for maintaining intraocular pressure?
* a. vitreous body, aqueous humor
b. iris, sclera
c. sclera, lens
d. lens, retina

45. Which is a structure of the macula that is made of calcium carbonate?

a. spiral organ
b. kinocilium
* c. otolith
d. tectorial membrane

The spiral organ is also known as the organ of Corti, which is the hearing organ. The spiral organ is located in the cochlea. The tectorial membrane is located right on the spiral organ's hair cells.

46. All of the following are components of the organ of Corti except:
a. outer hair cells
b. inner hair cells
c. supporting cells
* d. saccule

47. The spiral organ is located on top of which of the following structures of the inner ear?
* a. basilar membrane
b. tectorial membrane
c. saccule
d. utricle

The basilar membrane also divides scala tympani from the scala media.

48. Which of the following is a chamber of the cochlea that contains endolymph?
a. scala tympani
b. scala vestibuli
* c. scala media
d. utricle

49. The bony, inner, central part of the cochlea surrounded by the scala media, scala vestibuli and scala tympani are referred to as the:
a. macula
b. saccule
c. semicircular canal
* d. modiolus

The saccule, utricle and semicircular canals function in giving us equilibrium. The saccule and utricle contain a macula. Semicircular ducts contain cristae. Both the macula and cristae have hair cells that function in

dynamic equilibrium.

50. All of the following are chambers of the cochlea except:
a. scala tympani
b. scala vestibuli
c. scala media
* d. tympanic antrum

51. Which of the following is a not a structure of the middle ear?
a. stapedius muscle
b. tensor tympani muscle
* c. modiolus
d. incus

The modiolus is found in the cochlea in the inner ear. Other structures of the middle ear include the tympanic antrum, Eustachian tube, malleus, incus and stapes.

52. Which of the following contains perilymph and are structures of the inner ear?
* a. scala vestibuli, scala tympani
b. incus, malleus
c. stapes, incus
d. scala media, membranous labyrinth

53. Which of the following types of papillae always contain taste buds?
a. fungiform
* b. circumvallate
c. filiform
d. all of the above

Circumvallate papillae are found at the back of the tongue. Filiform papillae seldom contain taste buds. They are usually found at the anterior one-third to two-thirds of the tongue. The majority of fungiform papillae contain taste buds. They are usually located at the anterior tip of the tongue and the lateral aspects of the tongue.

54. The tympanic cavity is located in which bone of the skull?
a. frontal bone
* b. temporal bone

c. occipital bone
d. mastoid bone

55. The least abundant type of papillae that all have taste buds is identified as:

a. epiglottis
b. filiform
* c. circumvallate
d. fungiform

Filiform papillae usually have no taste buds. Most fungiform papillae have taste buds. The epiglottis does not have papillae, which are tongue projections or elevations on the exterior portion of the tongue.

56. Ampullary nerves are branches of which of the following cranial nerves?

a. cranial nerve 2
b. cranial nerve 3
c. cranial nerve 7
* d. cranial nerve 8

57. Circumvallate papillae have taste buds that are most associated with the taste of:

* a. bitter
b. salt
c. sweet
d. dryness

The taste of bitter is made possible by cranial nerve 9. The taste of salt and sweet are made possible by cranial nerve 7.

58. Which is a type of papillae that are considered the largest on the tongue?

a. filiform
b. fungiform
c. basal cell
* d. circumvallate

59. Which of the following is considered the central portion of the bony labyrinth?

a. basilar membrane
b. cochlea
* c. vestibule
d. round window

The middle ear is usually considered to be all the structures including and between the tympanic membrane, round window and oval window. The vestibule contains the utricle and saccule.

60. Which of the following structures contains both rods and cones?
a. sclera
b. lens
* c. retina
d. iris

61. Which of the following is a structure of the retina that is used for black and white vision?
* a. rods
b. sclera
c. cones
d. macula lutea

Cones are for color vision. The sclera is the white portion of the eye. The macula lutea has a very high quantity of cones. The optic disc has no cones and rods. The optic disc is also referred to as a blind spot.

62. Which of the following is associated with accomodation?
a. contraction of lens
b. constriction of pupil
c. convergence of eyes
* d. all of the above

63. Which of the following cranial nerves is most likely to be injured if a person has a noticeable drooping eyelid?
a. cranial nerve 2
* b. cranial nerve 3
c. cranial nerve 4
d. cranial nerve 8

A condition called ptosis is characterized as drooping of the eyelid. In such cases, cranial nerve 3 can be damaged. Cranial nerve 7 is responsible for

closing the eye. Cranial nerve 8 is for balance and equilibrium. Accomodation is made possible by cranial nerve 2 and 3.

64. Which is the correct visual pathway from first to last?

a. optic tract, optic nerve, optic chiasma, lateral geniculate nucleus of thalamus, cerebral cortex of occiptal lobe

b. optic nerve, optic tract, lateral geniculate nucleus of thalamus, optic chiasma, cerebral cortex of occipital lobe

* c. optic nerve, optic chiasma, optic tract, lateral geniculate nucleus of thalamus, optic radiations, cerebral cortex of occipital lobe

d. optic nerve, optic chiasma, optic tract, cerebral cortex of occipital lobe, lateral geniculate nucleus of thalamus

65. The vitreous chamber of the eye is located in which of the following locations?

a. iris to the cornea

b. lens to the iris

* c. retina to the lens

d. cornea to the iris

The anterior chamber and the posterior chamber make up the anterior cavity of the eye. The anterior chamber is found between the iris and cornea. The posterior chamber is located between the lens and iris. The vitreous body in the vitreous chamber has a significant part in maintaining intraocular pressure.

66. Why does the pupil constrict during accomodation?

* a. to prevent light from hitting the eye in the peripheral

b. to signal a fight-or-flight reaction

c. to increase one's awareness of his or her surroundings

d. to protect the lens from unwanted bacteria

67. Which structure of the ear is most medial?

a. malleus

b. pinna

c. tympanic membrane

* d. cochlea

The pinna is lateral to the tympanic membrane. The tympanic membrane is slightly lateral to the malleus. The malleus is lateral to the cochlea.

68. The thickening in the walls of the utricle and saccule is identified as the _________ and is a structure of the _________ ear.

a. ampulla, inner
* b. macula, inner
c. pinna, outer
d. external auditory meatus, outer

69. Which of the following is a kind of epithelial cell that makes up a taste bud?

a. basal cell
b. supporting cell
c. gustatory receptor cell
* d. all of the above

The opening portion of a taste bud is called the taste pore. There is gustatory hair surrounding the taste pore. Taste buds are found in projections of the tongue called papillae. The three types of papillae are called fungiform, circumvallate, and filiform. Specific portions of cranial nerve fibers that are responsible for taste are attached to the taste bud. Generally, cranial nerve 7 and cranial nerve 9 are responsible for the sense of taste.

70. Which is a type of papillae that usually contains taste buds and is found on the anterior tip of tongue?

a. circumvallate
* b. fungiform
c. filiform
d. supporting cell

71. Melanin is located in which area of the eye?

a. cornea
b. lens
* c. choroid
d. sclera

Melanin is necessary to act as a protection against glare. Without melanin in the choroid, it would be very hard to see anything clearly on a very bright sunny day. This is because there would be too much glare.

72. Which of the following structures surrounds a taste pore on a taste bud?

* a. gustatory hair
  b. basal cell
  c. macula
  d. cupula

73. All of the following are located in the bipolar cell layer of the retina except:
  a. horizontal cell
  b. amacrine cell
  c. bipolar cell
* d. rods

Rods and cones are located in the photorecepter layer of the retina. The deepest layer in the retina is the photorecepter layer. The bipolar cell layer lies more superficial compared to the photorecepter layer.

74. Which of the following layers of the retina is most superficial and closest to the retinal blood vessels?
  a. bipolar cell layer
  b. photoreceptor layer
* c. ganglion cell layer
  d. outer synaptic layer

75. Which is a portion of the macula lutea containing a large quantity of cones?
* a. central fovea
  b. anterior chamber
  c. aqueous humor
  d. optic disc

The optic disc does not have any cones or rods. This can be eliminated right away as a choice. The anterior chamber and posterior chamber make up the anterior cavity. The anterior cavity contains aqueous humor which is vital in maintaining specific levels of intraocular pressure. The aqueous humor is a liquid. It flows from the posterior chamber, to the anterior chamber and finally to the canal of Sclemm.

76. Light reaches which layer of the retina last?
  a. ganglion cell layer
* b. photoreceptor layer

c. bipolar cell layer
d. inner synaptic layer

77. The correct flow of aqueous humor from first to last is which of the following?
a. anterior chamber, posterior chamber, canal of Schlemm
b. posterior chamber, canal of Schlemm, anterior chamber
c. canal of Schlemm, anterior chamber, posterior chamber
* d. posterior chamber, anterior chamber, canal of Schlemm

Both the aqueous humor and the vitreous body contribute to maintaining intraocular pressure. The vitreous body also aids in keeping the retina and choroid connected together. The vitreous body is found in the vitreous chamber.

78. Which of the following is a false statement?
a. Rhodopsin is a pigment found in rods.
b. Rhodopsin absorbs green-blue light.
* c. Light enters the eye and never hits the retina since this can cause blindness.
d. The rhodopsin in the rods is responsible for humans perceiving shades of gray in dark surroundings.

79. The malleus, incus and stapes are structures of the middle ear and are considered:
a. tensor tympani muscles
* b. auditory ossicles
c. auditory tubes
d. semicircular canals

The tensor tympani muscle is attached to the malleus. The tensor tympani is important because it tenses in response to loud sounds. In this way, it aids in raising the tension of the tympanic membrane. This helps to dampen the vibration of a loud sound. The semicircular canals can be eliminated as a choice right away since they are structures of the inner ear.

80. Which of the following is not a structure of the inner ear?
a. ampulla
b. utricle
c. tectorial membrane

* d. tympanic antrum

81. Crista is found in which structure of the ear?
* a. ampulla of semicircular ducts
  b. cochlea
  c. tympanic antrum
  d. stapes

Crista is a projection found in the ampullae of semicircular ducts of the inner ear. The tympanic antrum and stapes are structures of the middle ear.

82. Which of the following is a jelly-like substance of the crista?
  a. kinocilium
  b. supporting cell
* c. cupula
  d. modiolus

83. Secretions of oil to lubricate the eyelids is made possible by which of the following?
  a. lacrimal apparatus
  b. lacrimal glands
* c. Meibomian glands
  d. ciliary process

The lacrimal glands are found over the lateral, superior portion of the eye. They secrete tears, which are called lacrimal fluid. Lacrimal fluid acts as a natural disinfectant for the eyes. It helps keep the eyes free of infection. Drinking enough water throughout the day also can help increase one's lacrimal fluid production. In this way, the eyes can be adequately hydrated. The lacrimal gland is located in the lacrimal fossa of the frontal bone. The lacrimal apparatus drains and creates tears.

84. All of the following are components of lacrimal fluid except:
  a. salt
  b. lysozyme
* c. adipose cells
  d. mucus

85. Which of the following is not a structure of the eye?
  a. lens

* b. chalazion
c. sclera
d. canal of Sclemm

A chalazion is not a normal structure of the eye; it is identified as a Meibomian gland cyst. Aqueous humor leaves the anterior chamber of the anterior cavity by way of the canal of Schlemm.

86. When a tear is released from the excretory lacrimal ducts, which is the correct order of travel from first to last?

a. lacrimal puncta, nasolacrimal duct, lacrimal canal
* b. lacrimal puncta, lacrimal canal, nasolacrimal duct
c. nasolacrimal duct, lacrimal puncta, lacrimal canal
d. lacrimal puncta, canal of Schlemm, lacrimal canal

87. Sebacious glands on the edges of the eyelids, which are not tarsal glands, are most closely identified as which of the following?

a. Meibomian glands
b. canal of Schlemm
c. lacrimal glands
* d. glands of Zeis

Meibomian glands are also sebaceous glands; they are found in the eyelids. They secrete oil through the edges of the eyelids. Meibomian glands are also known as tarsal glands.

88. Which of the following is not considered part of the vascular tunic?

a. ciliary body
b. choroid
* c. lens
d. iris

89. Which of the following glands produces mucus that hydrates the outer portion of the olfactory epithelium?

a. Meibomian glands
* b. Bowman's glands
c. lacrimal glands
d. glands of Zeis

The Meibomian glands, lacrimal glands, and glands of Zeis are glands of the eye. Basal cells, olfactory cells and supporting cells are found in olfactory

epithelium. Our sense of smell is made possible by the olfactory nerve, which is cranial nerve one.

90. Which of the following is responsible for black-and-white vision?
   a. sclera
   b. optic disc
* c. rods
   d. cones

91. Which of the following ligaments holds the lens of the eye in place?
* a. suspensory ligament
   b. lateral ligament
   c. superior ligament
   d. anterior ligament

There is a lateral, superior and anterior ligament of the malleus. The malleus is an auditory ossicle located in the middle ear. There are no blood vessels in the lens.

92. Which structure of the ear is covered by the secondary tympanic membrane?
   a. oval window
   b. external auditory meatus
* c. round window
   d. cochlea

93. Which of the following is a smooth muscle that changes the shape of the lens in order to see distant and close objects?
* a. ciliary muscle
   b. tensor tympani muscle
   c. stapedius muscle
   d. superior oblique muscle

The tensor tympanic muscle and tensor tympani muscle are found in the middle ear. The superior oblique muscle moves the eyeball down and in.

94. All of the following structures contain both perilymph and endolymph except:
   a. anterior semicircular canal
   b. lateral semicircular canal

c. posterior semicircular canal
* d. saccule

95. Which structure of the middle ear is considered the middle ossicle?
a. stapes
b. malleus
* c. incus
d. utricle

The malleus is connected to the incus. The incus is connected to the stapes. The malleus is hammer-shaped. The incus is anvil-shaped, and the stapes is stirrup-shaped. The tensor tympani muscle, stapedius muscle and tympanic antrum are all structures of the middle ear. The utricle is a structure of the inner ear. It is part of the membranous labyrinth and only contains endolymph.

96. The external auditory meatus is located within which bone of the skull?
a. parietal
b. mastoid
c. frontal
* d. temporal

97. The posterior and anterior chambers are portions of which of the following?
a. posterior cavity
* b. anterior cavity
c. lens
d. vitreous chamber

The vitreous body is found in the vitreous chamber, which is also known as the posterior cavity. The aqueous humor and the vitreous body are responsible for intraocular pressure in the eye. Aqueous humor is constantly replenished throughout the day; it plays a vital role in the creation of intraocular pressure.

98. Which of the following is considered an intrinsic eye muscle?
a. superior oblique muscle
* b. ciliary muscle
c. lateral rectus muscle
d. levator palpebrae